AWESOME DINOSAUR FACTS FOR KIDS

Discover Amazing Facts, Incredible Creatures, and the Wonders of the Prehistoric World!

John Hicks

FREE BONUS

SCAN TO GET OUR NEXT BOOK FOR FREE!

Table of Contents

INTRODUCTION

Welcome to the incredible world of dinosaurs! In this section, we'll travel back in time to discover the mysteries and wonders of these magnificent creatures that once roamed the Earth millions of years ago. From towering giants to tiny marvels, dinosaurs come in all shapes and sizes, each with their own unique features and behaviors.

Throughout our adventure, we'll meet some of the most famous dinosaurs that ever lived. From the mighty *Tyrannosaurus* to the horned *Triceratops*, we'll learn about their amazing adaptations, cool behaviors, and the environments they called home. However, our exploration won't stop there. We'll also uncover lesser-known dinosaurs such as the *Protoceratops*, *Maiasaura*, and *Compsognathus*, each with their own fascinating stories to tell.

As we journey through ancient landscapes, we'll discover how dinosaurs built their nests, took care of their babies, and found food to eat. We'll learn about their favorite snacks, like tasty plants and crunchy insects, and how they interacted with other dinosaurs. Along the way, we'll find out interesting facts about their bones, teeth, and the amazing places where scientists uncover their fossils.

However, our adventure isn't just about learning — it's also about using our imaginations and having fun. With illustrations, exciting stories, and interactive activities, we'll bring the world of dinosaurs to life.

In addition to learning about dinosaurs, we'll also discover the importance of protecting fossils, preserving habitats, and continuing to explore new dinosaur adventures. Throughout this book, we'll explore how scientists uncover dinosaur fossils, the importance of preserving their habitats, and how we can all help protect other incredible creatures and their ancient environments for future generations to enjoy.

WHY DINOSAURS ARE AWESOME

Before we explore different dinosaurs and their unique features, let's talk a little bit about why dinosaurs are so interesting. After all, these amazing creatures have captured the imaginations of people of all ages for hundreds of years. Here are some reasons why dinosaurs continue to amaze and delight us:

- **Gigantic size:** Imagine seeing a real-life giant like *Tyrannosaurus* or the long-necked *Brachiosaurus*.

- **Lots of different dinosaurs:** Did you know there are over 700 known kinds of dinosaurs and possibly many more that have yet to be discovered? That's a whole bunch of different shapes, sizes, and features to learn about.

- **Scientific names:** Compared to other creatures in modern times, dinosaurs like *Stegosaurus* and *Triceratops* are known by their scientific names. Looking up the translations can help you understand more about each animal or family group. For example, *Triceratops* translates to "three-horned face."

- **Finding fossils:** Think of dinosaur fossils like buried treasures waiting to be discovered. Learning about them helps us understand these incredible creatures that lived long ago.

- **Dinosaur behavior:** Did you ever wonder how dinosaurs hunted for food or took care of their babies?
- **Special skills:** Each dinosaur had special skills that helped them survive, such as sharp claws or super speed.
- **New discoveries:** Even scientists are still discovering new things about dinosaurs. Who knows what other secrets they might uncover?
  Nature's helpers: Dinosaurs played important roles in their habitats such as keeping plant populations in check.
- **Pop culture:** You've probably seen dinosaurs in movies, TV shows, books, and games.

As you can see, dinosaurs are a fascinating part of natural history. Whether you're interested in fossils or learning more about how ancient creatures lived in their day-to-day lives, this book is a great way to expand your knowledge of the dinosaur world.

HOW TO USE
THIS BOOK

You're going to learn a ton about dinosaurs throughout the following pages, but we don't want this to be a book you read once and never pick up again. Instead, we want it to be a resource that you'll use whenever you have a question about a dinosaur or you want to show other people what you know about these ancient animals.

Along with all the dinosaur facts you'll learn, there are also resources to help you continue your learning beyond these pages. We'll give you a list of museums, fossil parks, and tracksites to visit. If you can't take a trip right now, we'll provide you with other dinosaur activities and adventures that you can take advantage of right where you are.

So yes, it's time to read this book and learn all you can about dinosaurs. Once you're finished, keep it near at hand to refresh your memory, find other dinosaur resources, or plan your next prehistoric adventure. It's meant to be a book you come back to again and again.

CHAPTER ONE: WHAT ARE DINOSAURS?

DEFINING DINOSAURS

Before we dive into all the cool facts about dinosaurs, we're going to give you a definition that you can use to help identify these incredible creatures whenever you come across one in books, on television, or at a museum.

Dinosaurs were amazing beasts that lived on Earth millions of years ago. They weren't like any animals you see today, and they ruled the land long before humans even existed. Imagine a world where giant creatures roamed freely, some as tall as buildings and others as fierce as dragons.

One of the most fascinating things about dinosaurs was their size. Some dinosaurs were enormous, such as the mighty *Brachiosaurus*, which could reach up to 85 feet long and weigh as much as 50 tons. That's as long as six cars lined up bumper to bumper and as heavy as 10 elephants. Other dinosaurs, like the

Compsognathus, were only about the size of a chicken.

Additionally, dinosaurs had unique body structures with features like long necks, sharp claws, and bony plates that set them apart. Unlike most present-day reptiles, many dinosaurs were warm blooded, meaning they could regulate their body temperature like mammals.

Moreover, dinosaurs had a wide range of diets, from herbivores that fed on plants to carnivores that hunted other animals. Unlike mammals, dinosaurs laid eggs with hard shells. While some dinosaurs were swift and agile predators, others were slow-moving giants.

What's truly incredible about dinosaurs is that they lived all over the world. From hot, dry deserts to lush, green forests, dinosaurs adapted to many different habitats. Some preferred to live near rivers and lakes, while others roamed the open plains. They were truly masters of their environments.

You might wonder how we know so much about creatures that lived millions of years ago. Well, scientists have discovered dinosaur fossils all around the world. Fossils are the preserved remains or traces of ancient animals and plants. By studying these fossils, scientists have learned what dinosaurs looked like, how they moved, what they ate, and even how they behaved.

However, despite all that we've learned about dinosaurs, there's still so much we don't know. Some mysteries remain, like how dinosaurs went extinct. The most widely accepted theory is that a massive asteroid crashed into Earth about 65 million years ago, causing catastrophic changes to the environment. This event, along with other factors like volcanic eruptions and climate change, led to the extinction of the dinosaurs.

Some scientists believe in another theory. They think the number of dinosaurs might have slowly decreased as their environments changed over millions of years. Habitats shifted, food was harder to find, and dinosaurs needed to compete for resources to stay alive. Eventually, dinosaurs gradually went extinct, leaving behind a mystery.

Even though dinosaurs are no longer around, their legacy lives on. We can still learn from them and be amazed by their incredible feats. By studying dinosaurs, scientists can better understand the world we live in today and how it's changed over time.

WHEN DID DINOSAURS ROAM THE EARTH?

Dinosaurs lived during a time called the Mesozoic Era, which lasted from about 252 to 66 million years ago. Imagine a world without cars, buildings, or even people. Instead, vast forests covered the land, and huge oceans teemed with life. This was the perfect setting for dinosaurs to thrive. The Mesozoic Era is divided into three periods: the Triassic, Jurassic, and Cretaceous. Each of these periods had its own unique dinosaurs.

In the beginning, during the Triassic period, dinosaurs were just starting to appear. They shared the land with other creatures like early mammals and reptiles. These dinosaurs were small and not as mighty as the ones that came later, but they were the pioneers of their time.

During the Triassic period, which lasted from about 252 to 201 million years ago, the world looked very different than it does today. The landmasses were joined together in a supercontinent called *Pangaea*, surrounded by a single vast ocean known as *Panthalassa*.

As time marched on into the Jurassic period, dinosaurs began to dominate the landscape. This was the golden age of dinosaurs when massive creatures like the mighty *Brachiosaurus* and the fierce *Allosaurus* roamed the land.

The Jurassic period, spanning from approximately 201 to 145 million years ago, was characterized by lush forests and vegetation. Plants such as ferns, cycads, and conifers thrived during this time.

But the story doesn't end there. The Cretaceous period, from about 145 to 66 million years ago, brought even more incredible dinosaurs. This was the era of the fearsome *Tyrannosaurus*, the massive *Triceratops*, and the speedy *Velociraptor*.

The dinosaurs went extinct at the end of the Cretaceous period approximately 66 million years ago. The most popular theory is that dinosaurs were wiped out by a massive asteroid hitting the planet. The impact caused intense heat, wildfires, tsunamis, and a dust cloud that blocked out sunlight for months or even years. This catastrophe changed the environment and made it impossible for dinosaurs and other creatures to survive.

WHAT DINOSAURS ATE

Similar to humans and all other animals, the diet of dinosaurs was as varied as the dinosaurs themselves. Let's explore their prehistoric menu and discover what these ancient creatures liked to eat.

Plant-Eating Dinosaurs

Plant-eating dinosaurs are also known as *herbivores*. These gentle giants had a big appetite for plants and spent their days munching on leaves, ferns, and even entire trees. Some of the most famous plant-eating dinosaurs include:

- *Triceratops*: This horned dinosaur had a beak perfect for snipping leaves and branches. It loved to chomp on ferns and low-lying plants.
- *Stegosaurus*: With its taller back legs and shorter front legs, *Stegosaurus* most likely ate moss and shrubs that were low to the ground.
- *Brachiosaurus*: This massive dinosaur had a long neck and a big appetite. It used its height to reach the tops of trees, where it could feast on leaves and branches.

Meat-Eating Dinosaurs

Dinosaurs that only ate meat are known as *carnivores*. These fierce hunters roamed the land in search of prey, using their sharp teeth and claws to catch other dinosaurs. Some of the most famous meat-eating dinosaurs include:

- *Tyrannosaurus*: With its powerful jaws and teeth, *Tyrannosaurus* hunted large herbivores like *Triceratops*.
- *Velociraptor*: Despite its small size, the *Velociraptor* was a dangerous predator. It used its sharp claws and agility to hunt in packs, taking down much larger prey.
- *Allosaurus*: This fearsome predator had sharp teeth and a powerful bite. It hunted a variety of prey, from small dinosaurs to large plant-eaters like *Stegosaurus*.

Omnivorous Dinosaurs

Some dinosaurs were omnivores, meaning they ate both plants and meat. These adaptable dinosaurs had a varied diet, allowing them to thrive in a range of environments. Some examples of omnivorous dinosaurs are:

Gallimimus: This omnivore roamed the plains during the Late Cretaceous period. This speedy dinosaur likely fed on a variety of foods, including plants, insects, and small animals.

Oviraptor: This dinosaur from the Late Cretaceous period had a sharp beak for snipping plants, but it also had pointed teeth suitable for catching insects and other small prey.

THE FOUR MAIN TYPES OF DINOSAURS

No one knows for sure how many species of dinosaurs there were, but scientists have discovered and named about 700 species. However, experts suspect that there could have been as many as 160,000 species throughout the three eras when dinosaurs lived. Since we can't cover all 700 known dinosaurs in this book, we're going to narrow our selection down to the most well-known types.

Dinosaurs are divided into categories based on physical features. Those larger categories include smaller groups. A single dinosaur might fit into several different categories. This system helps scientists understand more about where each dinosaur fits out of all the dinosaurs that ever existed.

If that sounds complicated, just think of a modern-day animal like a dog. Out of plants and animals, a dog is considered an animal. From there, you could put it into another group of animals who have four legs. Within that category, a dog would belong in a smaller group for four-legged animals with tails. It's the same way for dinosaurs.

The first big split between dinosaurs is based on their hips. Saurischian dinosaurs have hips like reptiles, while ornithischian dinosaurs have hips like birds. After dividing dinosaurs into one of those two categories, they split again into four main groups.

Theropods

Theropods are known for having three toes with claws on each arm and foot. Most theropods were carnivores, but a few were also able to eat plants. Some of the deadliest predators fall into this category.

Tyrannosaurus undoubtedly struck fear with its massive jaws and powerful legs, while the feathered *Archaeopteryx* is considered one of the earliest birds. Other theropods include *Struthiomimus*, a toothless-beaked dinosaur that resembled modern ostriches.

Sauropods and Prosauropods

Imagine towering over the treetops, grazing on leaves with a long neck and a massive body. That's how sauropods and prosauropods lived. This category includes dinosaurs that eventually evolved to walk on all four legs. These gentle giants were some of the largest land animals to ever exist. *Brachiosaurus*, with its long neck and towering height, and *Diplodocus*, known for its whip-like tail, are two famous sauropods.

Cerapods

Cerapods are known as "horned-footed" dinosaurs. They were herbivores that had teeth designed for eating tough vegetation and plants. This larger group included ornithopods like *Iguanodon*, which had a horned beak and spiked thumbs that might have helped it search for food. Other cerapods include frilled dinosaurs such as *Triceratops* and pachycephalosaurs with thick, domed skulls.

Thyreophorans

Thyreophorans were herbivores that walked on all fours and typically had shorter legs in the front. They're famous for growing armor or bony plates to protect them from predators. *Stegosaurus* had bony plates along its back and a spiked tail. Meanwhile, *Ankylosaurus* was short and heavily armored with a club-like tail.

CHAPTER TWO:
MEET THE DINOSAURS

Now it's time to dive into the good stuff: learning about some of the most famous dinosaurs that have ever been discovered. As you read, keep in mind that this is far from a complete list. You'll have plenty to study on your own.

TYRANNOSAURUS: KING OF THE DINOSAURS

Meet the *Tyrannosaurus*! Its name translates to "tyrant lizard king," but it's also known as the king because it was such a formidable predator. This incredible creature roamed the Earth during the Late Cretaceous period.

An adult *Tyrannosaurus* was the size of a school bus and as heavy as four elephants put together. It had strong legs that allowed it to run up to 12 miles per hour while chasing its next meal. Scientists estimate that *Tyrannosaurus* ate enough meat every day to feed 80 people.

Even though paleontologists have figured out so much about dinosaurs, there are still a few mysteries. As you probably already know,

Tyrannosaurus had tiny arms compared to the rest of its body. Researchers aren't sure what exactly they were used for. It might have used its arms to stand up, attract a mate, or push other dinosaurs over while they were sleeping.

TRICERATOPS: THE HORNED WARRIOR

Let's turn our attention to the amazing *Triceratops*. This three-horned dinosaur had one short horn on its nose and two long ones above its eyes that it used to defend itself against predators.

Not every *Triceratops* had the same size horns, but the two long ones could grow to be over three feet, while its nose horn was around a foot long. *Triceratops* also had a bony frill at the back of its head to protect its neck from bites.

Even though *Triceratops* didn't have sharp teeth or claws, it wasn't an easy snack for predators. An adult could weigh over four tons and grow to be longer than a school bus. Despite its weight,

Triceratops could run up to 20 miles per hour, which is almost as fast as a modern elephant.

STEGOSAURUS: THE SPIKY DEFENDER

Meet the magnificent *Stegosaurus*. Its name means "roof lizard" in Greek because of the bony plates along its back. These plates weren't just for looks; they helped regulate the dinosaur's body temperature and may have even been used for defense. *Stegosaurus* also had a spiked tail that could be swung like a club. The spikes could grow up to two feet long.

Stegosaurus wasn't as big as some other dinosaurs. It measured around 30 feet long and stood about as tall as a basketball hoop. It ate leaves, ferns, and other vegetation, using its peg-like teeth to grind up tough plant material.

Stegosaurus fossils have been found all over the world. There's some evidence that *Stegosaurus* lived in groups, but researchers don't know for sure. Paleontologists hope to one day uncover fossilized nests, eggs, or hatchlings belonging to *Stegosaurus* to study them even more.

VELOCIRAPTOR: THE SWIFT HUNTER

The *Velociraptor* was fast, fierce, and full of surprises. Its name translates to "swift thief" in Latin. *Velociraptor* was one of the fastest dinosaurs around. It could sprint at speeds of up to 24 miles per hour, which makes it about a third as fast as a cheetah.

Velociraptor had a weapon hidden on each foot: a long, curved claw called the *sickle claw*. This claw could be used to slash at prey. Most of the time, *Velociraptor* hunted in packs to take down larger dinosaurs. Despite what you may have seen in movies, *Velociraptor* was only about two feet tall and six feet long, making it too small to hunt on its own. Scientists believe they could have used sounds such as chirps, clicks, growls, and squawks to coordinate.

Recent discoveries also suggest that *Velociraptor* might have been covered in feathers to stay warm or communicate with other dinosaurs. Having feathers on its feet could also have helped it grip the ground while running.

BRACHIOSAURUS: THE GENTLE GIANT

Introducing the *Brachiosaurus*! This magnificent creature was one of the largest land animals to ever walk the Earth. *Brachiosaurus* could grow as tall as 40 feet, and its neck alone could be 30 feet long. Its long neck allowed it to reach high into the treetops to munch on leaves that other dinosaurs couldn't reach.

To pump blood all the way up to its head, *Brachiosaurus* had a special heart with extra chambers. This helped keep its blood pressure stable despite its height, ensuring that it stayed healthy and strong.

Scientists used to think that *Brachiosaurus* lived in water because it would have been too heavy to move around on land. It also had nostrils on the top of its head. Researchers later realized *Brachiosaurus* wouldn't have been able to breathe underwater and changed their views on its habitat.

ANKYLOSAURUS: THE ARMORED TANK

Weighing up to 10,000 pounds, *Ankylosaurus* was a walking tank, equipped with heavy armor and powerful defenses. Thanks to its bony shell, it could withstand attacks from many different types of predators.

Some scientists believe that the bony plates on the Ankylosaurus's back could have been shed and replaced over time, like how some reptiles shed their skin. This regenerative ability would have allowed the *Ankylosaurus* to maintain its protective armor throughout its life.

One of the most impressive features of the *Ankylosaurus* was its tail club. This massive weapon was made of solid bone and could be swung to defend against attackers. *Ankylosaurus* may also have dug trenches to hide from predators or search for roots to eat.

PARASAUROLOPHUS: THE CRESTED DINOSAUR

Now, we're going to examine the *Parasaurolophus*. This duck-billed dinosaur was named for the long, hollow crest on top of its head, which resembles the shape of a tube. Scientists believe that the crest of the *Parasaurolophus* may have been used to produce sounds like a trumpet or trombone. Air could have been blown through the crest, creating different pitches and tones.

Parasaurolophus was an herbivore that ate leaves, twigs, and pine needles. Its jaw contained over 1,000 tiny teeth to help it break down whatever vegetation it could find. When it lost a tooth, another was always ready to take its place.

Parasaurolophus likely lived in herds. Being part of a herd provided protection against predators and allowed the members to interact. Scientists believe that *Parasaurolophus* may have migrated seasonally in search of food and suitable breeding grounds. These migrations could have taken them hundreds of miles across ancient landscapes.

SPINOSAURUS: THE RIVER DINO

Next, we'll take a look at *Spinosaurus*. This remarkable dinosaur had a long, sail-like structure on its back. It was one of the largest carnivorous dinosaurs to ever exist. It could grow up to 50 feet long, which is slightly longer than a semi-truck trailer.

Spinosaurus was well adapted to life in and around rivers and swamps. It had long, slender jaws filled with sharp teeth, perfect for catching fish and other aquatic prey. Its nostrils were positioned high on its skull, allowing it to breathe while mostly submerged in water.

Spinosaurus was also an excellent swimmer, thanks to its streamlined body, paddle-like feet, and long, powerful tail. It could glide through the water with ease, using its tail for propulsion like a modern-day crocodile.

The sail on the back of Spinosaurus wasn't just for show. It may have helped regulate its body temperature, acting like a giant solar panel to soak up the sun's warmth during the day and getting rid of extra heat at night.

ALLOSAURUS: THE BIG CLAW

Next up, we uncover the mysteries of the Allosaurus. The name Allosaurus means "different lizard" in Greek. It was given this name because its bones were different from other dinosaurs that had already been discovered at that time.

Allosaurus had large, sharp claws that it used to grasp and hold onto prey during a chase. Despite its large size, it was a fast runner that could sprint at speeds of up to 30 miles per hour. Its forward-facing eyes and excellent vision also gave it an advantage while hunting.

Fossil evidence shows that *Allosaurus* crushed bones and may have scavenged carcasses left behind by other predators. It wasn't a picky eater. Some scientists believe *Allosaurus* may have also hunted in packs to catch larger dinosaurs.

DIPLODOCUS: THE LONG-NECKED GIANT

Diplodocus was an extremely large herbivore that lived during the Late Jurassic period. Its name means "double-beamed" in Greek since it had double-beamed chevron bones in its tail. These bones gave its tail extra strength and flexibility. It most likely used its tail as a whip to defend against predators.

Diplodocus was one of the longest dinosaurs to ever walk the Earth. It could reach lengths of up to 90 feet once it finished growing. An adult *Diplodocus* didn't reach full size until it was about 20 years old.

Diplodocus fossils have even played a role in diplomatic relations. Andrew Carnegie funded a dig when dinosaur bones were discovered in Wyoming in 1898. Those bones made up the first *Diplodocus* skeleton ever found.

King Edward VII was fascinated by the dinosaur and wanted to know if there were any others to display in London. Since there weren't any other *Diplodocus* fossils, the Carnegie Museum of Natural History in Pittsburgh made plaster copies of the skeleton and shipped them to London. Other copies were sent to Germany, Austria, France, Spain, Russia, and Argentina.

ARCHAEOPTERYX: THE FEATHERED DINO-BIRD

Let's return to the skies of the ancient world and explore the bird-like *Archaeopteryx*. The name *Archaeopteryx* means "ancient wing" in Greek. Its fossilized remains provide strong evidence of the link between dinosaurs and birds, making it a crucial discovery in the study of evolution.

Archaeopteryx had many features in common with birds, including feathers, wings, and a wishbone. It also had hollow bones that made it lightweight and suited for flight. When its fossils were first discovered, many scientists argued that *Archaeopteryx* was the oldest known bird instead of a dinosaur.

Even though *Archaeopteryx* had wings and feathers, it most likely glided from tree to tree rather than flying long distances. However, researchers are limited by the low number of

35

Archaeopteryx fossils. So far, only 12 sets of fossils have ever been found.

IGUANODON: THE THUMB-SPIKED DINOSAUR

Iguanodon was one of the first dinosaurs to be discovered. At first, paleontologists thought the *Iguanodon* tooth found in England in 1822 belonged to a gigantic version of iguanas. Over time, people realized dinosaurs were entirely different creatures.

We now know that *Iguanodon* walked on two legs, but it could use all four legs when it was moving at slower speeds. It was roughly the size of a modern-day African elephant, making it one of the largest herbivores of its time.

Most of the bones in its skull were flexible. This made it easier for *Iguanodon* to chew tough plants. Its ridged teeth also helped it grind up ferns and other plants that grew by water.

CARNOTAURUS: THE MEAT-EATING BULL

Carnotaurus shares some similarities with *Tyrannosaurus*, but they lived in separate parts of the world during the Cretaceous period. Its name means "meat-eating bull" in Latin since *Carnotaurus* had two short horns above its eyes.

Impressions of *Carnotaurus* skin have been found in fossils, revealing a rough texture like that of modern-day crocodiles. This bumpy skin may have provided protection against predators or helped regulate body temperature. Scientists also believe that *Carnotaurus* may have had colorful skin patterns or markings to help it blend into its surroundings while hunting.

Carnotaurus had incredibly powerful legs and a tail that was designed to help it run faster. It could reach speeds of up to 35 miles per hour, almost triple the top speed of *Tyrannosaurus*.

MICRORAPTOR: THE TINY AVIATOR

The name *Microraptor* means "tiny thief" in Greek. In fact, it's the smallest of all dinosaurs ever discovered. An adult *Microraptor* grew to be about a foot tall, and it only weighed two pounds.

Microraptor had wings on all four limbs. This unique adaptation allowed it to glide through the air like a modern-day flying squirrel. It also had a tail with feathers that might have helped it steer while hunting lizards, insects, and other small prey.

Microraptor had shiny feathers that it might have used to attract a mate. Peacocks and hummingbirds have feathers with the same type of metallic look.

APATOSAURUS: THE DECEPTIVE LIZARD

The *Apatosaurus* was a humongous dinosaur that was initially confused with *Brontosaurus*. That's why its name translates to "deceptive lizard" in Greek. It's now known that *Apatosaurus* was much larger than any other dinosaurs with similar skeletons.

Apatosaurus reached lengths of up to 75 feet and weighed as much as 25 tons. Its neck could grow as long as 30 feet. Compared to other large dinosaurs, *Apatosaurus* grew quickly and could reach full size in 10 years.

Apatosaurus laid round eggs that could be up to a foot wide. Several groups of eggs have been found. Instead of nests, *Apatosaurus* laid eggs in a line. Scientists think this could mean that *Apatosaurus* didn't take care of their eggs or protect their hatchlings.

UTAHRAPTOR: THE UTAH THIEF

The *Utahraptor* was the largest known raptor, reaching lengths of up to 23 feet and weighing as much as 1,000 pounds. Like other raptors, *Utahraptor* likely had feathers covering its body.

Utahraptor was a relatively fast runner, capable of sprinting at speeds of up to 25 miles per hour. Compared to other raptors, it wasn't able to jump or pounce as well because of its larger size.

Before the *Utahraptor* was discovered, scientists thought all raptors were smaller predators from the Late Cretaceous. *Utahraptor*, however, lived during the Early Cretaceous and grew much larger than many other species in the same family.

EDMONTOSAURUS: THE DUCK-BILLED DINO

Here's a lesser-known dinosaur that deserves some attention. *Edmontosaurus* was one of the largest dinosaurs of the duck-billed dinosaurs, reaching lengths of up to 43 feet and weighing almost as much as 7,700 pounds.

It was an herbivore that used its duck-like bill to strip leaves and branches from trees. *Edmontosaurus* didn't have teeth like those of modern ducks. Instead, it had hundreds of small, closely packed teeth that were perfect for grinding up food. It mostly ate twigs, pine needles, and cones.

It wasn't very fast and had few defenses to protect itself from predators. However, *Edmontosaurus* may have had unusually strong senses that helped it know when danger was near.

BRONTOSAURUS: THE THUNDER LIZARD

Brontosaurus means "thunder lizard" in Greek. It was given this name because of its massive size and the ground-shaking footsteps it may have made. *Brontosaurus* was one of the largest dinosaurs to ever roam the Earth. It reached lengths of up to 75 feet and weighed as much as 33 tons.

Brontosaurus had a series of bony spines running along its back. It also had claws that it might have used to hold onto trees, dig for water, or build nests. No *Brontosaurus* nests have been found to date, so scientists are still learning about how they might have raised their young.

Some paleontologists believe that *Brontosaurus* used its neck to fight and defend itself. It had a stronger neck than other dinosaurs, and modern-day animals like giraffes use their necks in similar ways.

GALLIMIMUS: THE CHICKEN MIMIC

Even though the *Gallimimus* is known as the "chicken mimic," it didn't really resemble a chicken. It was much larger than a chicken and probably looked more like the *Velociraptor* than most birds we're familiar with.

Gallimimus was one of the fastest dinosaurs, capable of running at speeds of up to 34 miles per hour. Its long legs and lightweight build allowed it to outrun most predators. It reached lengths of up to 20 feet and weighed around 450 pounds.

Gallimimus was omnivorous. It ate ferns and other vegetation dropped by larger dinosaurs. It also ate eggs and small animals such as lizards. Based on the design of its beak, it swallowed its prey whole.

STEGOCERAS: THE ROOF-HORNED DINO

Even though its name is similar, the *Stegoceras* shouldn't be confused with the *Stegosaurus*. The name *Stegoceras* means "roof horn" in Greek. It was given this name because of the bony dome on its skull.

Stegoceras was a small dinosaur, reaching lengths of up to 8 feet and weighing up to 100 pounds. It had strong legs and excellent balance. Scientists used to think this was to help it stay upright while butting heads with other dinosaurs or members of its species.

However, some paleontologists think its skull was too soft to use as a weapon. *Stegoceras* skulls do seem to lack any signs of breaks or damage. More research is needed to understand exactly why *Stegoceras* had a bony dome on top of its head.

OVIRAPTOR: THE EGG THIEF

Next, let's learn about the *Oviraptor*, also known as the "egg thief." *Oviraptor* has a bit of an unfair reputation. It probably didn't eat eggs at all. Based on its jaw, its diet likely consisted of clams, insects, and seeds.

The first *Oviraptor* fossils were discovered near a nest of dinosaur eggs, leading scientists to believe that it had been trying to steal them. In reality, the *Oviraptor* had been guarding its own nest.

This shows that *Oviraptor* brooded their eggs just like modern-day birds. Paleontologists only figured out their mistake when they realized one of the eggs contained a baby *Oviraptor*.

DEINONYCHUS: THE TERRIBLE CLAW

The name *Deinonychus* means "terrible claw" in Greek because of the large, curved claw on each of its hind feet. When *Deinonychus* didn't need the claw, it held it off the ground while walking to keep it as sharp as possible.

Deinonychus was a carnivore that likely preyed on small dinosaurs and other animals during the Early Cretaceous period. Paleontologists believe that it was a pack hunter that used its speed and intelligence to take down larger prey.

Deinonychus also had several bird-like features, including feathers and short wings. Even its respiratory system and digestion were similar to modern-day birds. The discovery of *Deinonychus* fossils in the 1960s finally convinced many scientists that dinosaurs and birds are related.

COMPSOGNATHUS: THE ELEGANT JAW

Compsognathus once held the record for being the smallest dinosaur. An adult grew to about 3 feet long and weighed up to 12 pounds. *Compsognathus* also had a slender jaw and three-toed feet.

Compsognathus was a carnivore that preyed on insects and small lizards. A *Compsognathus* fossil found in Germany even had the skeleton of a lizard still inside its rib cage where its stomach would have been. While hunting, it used its long tail to maintain balance and make sharp turns.

Because of its small size and delicate bones, *Compsognathus* fossils are rare. Only two sets have ever been found: one in Germany and one in France.

PROTOCERATOPS: THE FIRST HORNED FACE

The name *Protoceratops* means "first horned face" in Greek. It was given this name because it was one of the first dinosaurs of its kind discovered by scientists. *Protoceratops* was a medium-sized herbivore, reaching lengths of up to 8 feet and weighing as much as 400 pounds.

Protoceratops belongs to the same family as other horned dinosaurs such as *Triceratops* and *Styracosaurus*. However, unlike its larger relatives, *Protoceratops* had a horned beak instead of horns above its eyes or on its nose.

Hundreds of *Protoceratops* fossils have been found around the world. This includes younger *Protoceratops* and new hatchlings in nests. These skeletons made it easier to understand how *Protoceratops* grew over time.

MAIASAURA: THE GOOD MOTHER LIZARD

We've come to the last dinosaur on our list, and while it's not one of the more famous dinosaurs, it deserves its place because of its parenting habits. The *Maiasaura*, or "good mother lizard," took care of its young for longer than many other dinosaurs.

After studying fossilized nests and the bones of hatchlings, researchers believe *Maiasaura* babies had unusually soft bones. They probably stayed in their nests for up to a month while their skeletons grew stronger. Adults brought them food and kept them safe from predators. *Maiasaura* nested together in colonies to make this easier.

Maiasaura had a duck bill and teeth. It ate leaves, berries, and seeds as its main diet. At times, *Maiasaura* would also eat wood and shells.

Eating wood might have allowed it to survive during droughts or when food was hard to find.

CHAPTER THREE: DINOSAUR HABITATS

MESOZOIC ERA ENVIRONMENTS

The Mesozoic Era lasted from about 252 to 66 million years ago. During this time, dinosaurs roamed the land, gigantic reptiles ruled the seas, and flying reptiles filled the skies.

Triassic Period

In the early Triassic, the climate was generally hot and dry, with large desert regions dominating some areas. However, there were also lush forests near water sources where plants and early dinosaurs thrived. The Triassic Period lasted until about 201 million years ago.

Jurassic Period

During the Jurassic Period, the climate became warmer and wetter. Tropical forests and swamps were common. Dinosaurs thrived during this time. Rivers and lakes teemed with life, including ancient crocodiles and fish. This period ended around 145 million years ago.

Cretaceous Period

In the Cretaceous Period, the Earth's climate continued to change. The supercontinent Pangaea began to break apart and create the continents we know today. Tropical forests still covered a lot of the land, but there were also grassy areas and other plants.

Marine Environments

The Mesozoic seas were full of life. Enormous marine reptiles like *Plesiosaurus* ruled the oceans, while coral reefs flourished in shallow waters. Giant ammonites, relatives of modern squids and octopuses, and ancient sharks also lived during this time.

Sky High

The skies of the Mesozoic were filled with flying reptiles known as pterosaurs. These incredible creatures had wingspans ranging from a few inches to over 30 feet. They soared above the land and sea, hunting for fish or scavenging for food. Some pterosaurs even had crests on their heads that they might have used to communicate.

DIFFERENT TYPES OF HABITATS

Dinosaurs lived in all parts of the world, which means they lived in all types of habitats. Each dinosaur had features that made it able to survive in its preferred area. As mentioned before, there were thousands of different dinosaur species that adapted to these environments.

Forests

Picture yourself walking through a dense, green forest filled with towering trees and lush vegetation. In these leafy havens, plant-eating dinosaurs like *Triceratops* and *Stegosaurus* munched on leaves, ferns, and other tasty plants. Meanwhile, carnivorous dinosaurs like *Velociraptor* and *Tyrannosaurus* prowled through the undergrowth, hunting for their next meal.

Grasslands

Imagine wide-open spaces covered in tall grasses swaying gently in the breeze. Grasslands were another common habitat for dinosaurs, especially during the later parts of the Mesozoic Era. These areas were home to large herds of plant-eating dinosaurs such as *Ankylosaurus*. These vast plains were also home to speedy predators like *Deinonychus* that relied on speed to catch their prey.

Wetlands

Wetlands and marshes were full of life during the time of the dinosaurs. Here, you would find long-necked dinosaurs wading through shallow waters, using their height to reach plants and trees growing along the banks. Crocodile-like dinosaurs such as *Spinosaurus* lurked in the murky waters, waiting for an unsuspecting meal to pass by.

Mountains

Mountains weren't the most popular place for dinosaurs to live, but some did adapt to the cooler temperatures. Scientists believe that *Edmontonia* and other types of armored dinosaurs might have lived in the mountains.

Deserts

Deserts weren't as common during the Mesozoic Era as they are today, but they still existed in certain regions. Here, you might find dinosaurs like the *Protoceratops* that adapted to survive in hot environments with limited water and vegetation.

Coastal Areas

Coastal areas where land meets the sea were vibrant habitats for dinosaurs. *Oviraptor* lived on the shores, while powerful *Spinosaurus* hunted both on land and in water. Coastal zones served as nesting grounds for dinosaurs that kept their eggs hidden in sandy nests.

Regardless of where they lived, dinosaurs were masters of adaptation. From dense forests to open grasslands, from swampy wetlands to rugged mountains, dinosaurs conquered every corner of the Earth during their reign. By studying the habitats where dinosaurs lived, scientists can learn more about how these incredible creatures survived and evolved over millions of years.

CHAPTER FOUR:
HOW SCIENTISTS
STUDY DINOSAURS

PALEONTOLOGY: THE SCIENCE OF DINOSAURS

The study of dinosaurs is called *paleontology*, and it's as exciting as you might imagine. If you've ever wondered what it would be like to discover a dinosaur fossil that's millions of years old, then it's time to read on and learn more about this fascinating job.

What Is Paleontology?

Paleontology is like being a detective of the past. It's the scientific study of ancient life, including dinosaurs, plants, insects, and even early humans. Paleontologists use clues from fossils to piece together what life was like millions of years ago.

How Do Paleontologists Study Fossils?

Once fossils are discovered, paleontologists carefully excavate them from the ground using tools such as brushes, picks, and shovels. They record important details about where the fossils were found and how they were positioned in the layers of rock.

Back in the lab, paleontologists clean, study, and analyze the fossils. They might use special techniques like CT scans and 3D imaging to get a closer look inside the fossils without damaging them.

By comparing fossils from different places and times, paleontologists can learn how dinosaurs evolved over millions of years and how they were related to each other.

What Have Paleontologists Learned?

Paleontology is an incredible field that allows us to travel back in time and explore the ancient world of dinosaurs. By studying fossils, paleontologists can unlock the secrets of the past and help us understand how life on Earth has changed over millions of years. Thanks to the work of paleontologists, we've learned so much about dinosaurs and the world they lived in.

It all started on February 20, 1824, when William Buckland introduced *Megalosaurus* to the Geological Society of London. It was the first dinosaur to ever be discovered and scientifically studied. The word *dinosaur* wasn't invented until 18 years later.

In the late 1800s, paleontologists Othniel Charles Marsh and Edward Drinker Cope became rivals. They fought constantly and even destroyed each other's work. In total, they documented over 136 different types of dinosaurs.

In 1923, researchers from the American Museum of Natural History found the first dinosaur eggs while digging in Mongolia. This discovery made people curious about how dinosaurs laid eggs and took care of their young.

Finally, in the 1990s, paleontologists discovered *Sinosauropteryx* in China. The fossil included impressions that looked feathers. It was the first time scientists had ever found proof that dinosaurs had feathers even if they weren't closely related to birds.

FOSSILS AND EXCAVATION

Now we're going to look a little deeper into fossils and the process of uncovering them through excavation. Some paleontologists love to go digging for fossils in the field. Others focus on conducting research, taking care of fossils in museums, and analyzing specimens in a laboratory.

How Are Fossils Formed?

Fossils form in a few different ways. Sometimes, when an animal dies, its bones are buried under layers of dirt or mud. Over time, the bones can become fossilized as minerals replace the original bone material. Other times, footprints or imprints of plants and animals can be preserved in rock or mud, leaving behind a record of their existence.

Types of Fossils

Fossils come in many shapes and sizes, each telling its own story about life in the past. Here are some common types of fossils:

- **Bones and teeth:** These are the most common types of fossils. They can tell us a lot about the size, shape, and behavior of ancient animals.
- **Footprints:** Footprint fossils provide clues about how dinosaurs moved and interacted with their environment.
- **Eggs and nests:** Fossilized eggs and nests help us understand how dinosaurs reproduced and cared for their young.
- **Coprolites:** Coprolites are fossilized poop. Yeah, you read that right! These fossils can reveal what dinosaurs ate and how their digestive systems worked.

Excavating Fossils

Excavation is the process of carefully digging up fossils from the ground. First, paleontologists need to find fossils. They might search for clues in places where rocks are exposed, like cliffs or canyons. Sometimes, fossils are discovered by accident, like when a construction crew digs up a dinosaur bone while building a road.

Once fossils are found, it's time to start digging. However, this isn't like digging for buried treasure in your back yard. Paleontologists use special tools such as brushes, picks, and even dental tools to carefully remove dirt and rock from around the fossils. They work slowly and gently to avoid damaging the fossils.

As they dig, paleontologists record important details about where the fossils were found and how they were positioned in the ground. This information helps scientists piece together the story of the ancient world.

Once the fossils are fully excavated, they need to be carefully transported back to the lab for further study. This might involve wrapping them in protective material or placing them in special containers to keep them safe during transit.

Tools of the Paleontologist Trade

Paleontologists use a variety of tools to excavate fossils and study them in the lab. Here are some of the most important tools in a paleontologist's toolkit:

- **Brushes:** Soft brushes are used to gently sweep away dirt and debris from around fossils without damaging them.
- **Picks and chisels:** These tools are used to carefully chip away rock and sediment.
- **Air pens:** These small devices hammer away at the surrounding rock to uncover fossils. Since an air pen is powered by compressed air, it can strike the rock thousands of times per minute.
- **Dental picks:** Like the tools your dentist uses, dental picks are perfect for delicately removing stubborn bits of rock.
- **Sieves:** Sieves are used to sift through sediment to find tiny fossils, like teeth or bone fragments, that might otherwise be overlooked.

Why Do We Excavate Fossils?

Excavating fossils is like uncovering pieces of a puzzle. Each fossil tells a story about ancient life and helps scientists understand how plants and animals have evolved over millions of years. By studying fossils, we can learn about the past and better understand the world we live in today.

The next time you see a dinosaur skeleton in a museum, remember that it's not just a bunch of bones. It's a window into the past and a reminder of the incredible creatures that once roamed the Earth.

RECONSTRUCTION OF DINOSAURS

Paleontologists are also tasked with taking fossils and reconstructing the animal they came from. In most cases, they can't find all the bones that belonged to one animal, so they must use their knowledge of dinosaurs to fill in the blanks. This means that some reconstructions have to be redone when scientists gain knowledge from newly discovered fossils.

Whether they're assembling a new dinosaur or reassembling one that needs to be updated, scientists follow certain procedures to make sure their models are as accurate as possible.

What Is Dinosaur Reconstruction?

Dinosaur reconstruction is like solving a big prehistoric puzzle. It's the process of piecing together what dinosaurs looked like based on the fossils they've left behind. Like building a model or drawing a picture, paleoartists use their imagination and scientific knowledge to create realistic depictions of dinosaurs.

Finding Clues in Fossils

The first step in reconstructing dinosaurs is studying their fossils. Bones, teeth, footprints, and even skin impressions can provide important information about a dinosaur's size, shape, and features.

Using Comparative Anatomy

Paleontologists compare dinosaur fossils to the skeletons of animals that are alive today, like birds and reptiles, to help figure out how dinosaurs moved and what they might have looked like. For example, a dinosaur's leg bone might be similar in shape to a bird's, suggesting that the dinosaur was a fast runner.

Considering Skin, Feathers, and Scales

Once paleontologists have an idea of a dinosaur's body shape and size, they can start thinking about its skin, feathers, or scales. Some dinosaurs had scaly skin, while others had feathers. Fossilized skin impressions and studies of living animals help paleoartists make educated guesses about a dinosaur's outer appearance.

Recreating the Face

One of the most challenging parts of reconstructing a dinosaur is figuring out what its face looked like. Since soft tissues like muscles and skin don't usually fossilize, paleoartists must use their imagination and scientific evidence to sculpt the face of a dinosaur. They might look at the shape of the skull, the size of the eye sockets, and the position of the nostrils to make an educated guess.

Bringing Dinosaurs to Life

Once all the pieces of the puzzle are put together, it's time to bring the dinosaur to life! Paleoartists use their artistic skills to create drawings, paintings, sculptures, or even computer-generated models of what they think the dinosaur looked like. They might add details such as color patterns, texture, and movement to make their reconstructions as lifelike as possible.

Challenges of Reconstruction

Reconstructing dinosaurs isn't always easy. Sometimes, there's not enough fossil evidence to accurately piece together a complete picture of a dinosaur. Other times, scientists disagree about how certain features should be interpreted.

For example, when scientists found the spiky thumb bone of an *Iguanodon* fossil, they originally thought it belonged on the dinosaur's head. It took some time for them to realize it was actually a thumb that *Iguanodon* most likely used to eat fruit or protect itself from predators.

CHAPTER FIVE: FUN DINOSAUR FACTS

PTERODACTYLUS means "winged finger"

DINOSAUR COLORS AND PATTERNS

With only bones to look at, it takes some work to determine the colors and patterns of dinosaur skin, feathers, and fur. While we can't travel back in time to see them for ourselves, scientists have some clever ways to make educated guesses about the colors of dinosaurs.

Fossil Evidence

Believe it or not, fossils themselves can provide clues about a dinosaur's color. Sometimes, fossilized skin impressions are preserved along with the bones. These impressions show textures and patterns that can help scientists imagine what the skin of a dinosaur might have looked like.

Comparing to Living Animals

Paleontologists also compare dinosaurs to modern animals to make guesses about their colors. For example, if a dinosaur is closely related to birds, it might have had feathers that were similar in color to modern birds.

Pigment Cells

Recent scientific discoveries have shown that dinosaurs might have had pigment cells called *melanosomes* in their skin, feathers, or scales. By studying the shape and arrangement of melanosomes in fossils, scientists can make educated guesses about the colors of dinosaurs. For example, round melanosomes might suggest dark colors like black or brown, while rod-shaped melanosomes might suggest bright colors like red or yellow.

Patterns in Nature

Like how animals today use colors and patterns for camouflage, communication, or attracting mates, dinosaurs likely had their own reasons for being colorful. Some dinosaurs might have had stripes or spots to help them blend in with their surroundings, while others might have had bright colors to attract mates or intimidate rivals.

Examples of Dinosaur Colors and Patterns

Now, let's look at some examples of dinosaur colors and patterns based on scientific evidence and artistic interpretations.

- *Tyrannosaurus*: The mighty *Tyrannosaurus* might have had scaly skin with earthy tones like green, brown, or gray, helping it blend into its forest or swamp habitat. Some scientists also think *Tyrannosaurus* might have had stripes or patches of color for camouflage.
- *Triceratops*: This horned dinosaur might have had a colorful frill around its neck, with patterns of stripes or spots to help it communicate with other *Triceratops* or attract mates.
- *Velociraptor*: With its feathers and bird-like appearance, *Velociraptor* might have had colorful feathers like modern birds. Some scientists think *Velociraptor* might have had stripes or patterns on its feathers for camouflage or display.

- *Stegosaurus*: This armored dinosaur might have had patches of color on its plates and spikes. Patterns may have helped it regulate its body temperature or communicate with another *Stegosaurus*.
- *Ankylosaurus*: With its thick armor plating and bony club tail, *Ankylosaurus* was a formidable herbivore. Its armor might have been a muted shade of gray or brown, providing camouflage against predators. Some scientists believe that its armor plates and club tail could have had contrasting patterns, such as bands or stripes, which may have served as a warning to potential attackers or as visual signals within its own species.

DINOSAUR SOUNDS AND COMMUNICATION

Have you ever wondered what sounds dinosaurs made? Well, just like with colors and patterns, we can't travel back in time to listen to them. However, scientists have come up with ways to guess what dinosaur noises might have been like.

Vocal Structures

Like birds and reptiles today, dinosaurs likely used their lungs, vocal cords, and air sacs to make sounds. The size and shape of a dinosaur's vocal structures can give scientists clues about what kinds of sounds it might have been capable of making.

Comparing to Living Animals

Paleontologists also compare dinosaurs to modern animals to guess what sounds they might have made. For example, if a dinosaur is closely related to birds, it might have made chirping or squawking noises.

Body Language

Dinosaurs also likely used body language to communicate with each other. For example, a large predator such *Tyrannosaurus* might have stomped its feet, bared its teeth, or flapped its arms to show dominance or attract a mate.

78

Examples of Dinosaur Sounds and Communication

Let's take a closer look at some examples of dinosaur sounds and communication based on scientific evidence and artistic interpretations.

- **Tyrannosaurus**: *Tyrannosaurus* might have made deep, booming roars to communicate. These roars could have been used to establish territory, attract mates, or intimidate rivals. Some scientists also think the *Tyrannosaurus* might have made softer sounds like grunts or hisses to communicate up close.

- **Stegosaurus**: This ancient, armored dinosaur possibly made low rumbling noises by shaking its bony plates, perhaps to communicate within its herd or as a warning signal to predators.

- **Triceratops**: *Triceratops* may have made deep grunts or bellows to find mates or defend its territory against rivals.

- **Hadrosaurus**: Recognized for its unique "duck-billed" appearance, *Hadrosaurus* likely produced trumpet-like calls or honks to communicate with its herd members and alert them to potential dangers in the marshy landscapes it inhabited.

- *Velociraptor*: This swift predator might have made high-pitched screeches or calls like modern-day birds, possibly for communication within its pack during hunts.

- *Brachiosaurus*: With its long neck and massive size, *Brachiosaurus* may have produced deep, resonating calls or bellows. These sounds could have been a form of long-distance communication or a way to attract potential mates during the breeding season.

- *Dilophosaurus*: Dilophosaurus likely made low-frequency rumbles or growls to protect its territory and attract mates.

- *Ankylosaurus*: With its heavily armored body and clubbed tail, Ankylosaurus might have made deep thumping sounds by striking its tail against the ground. This could serve as a warning signal to predators or a way of communicating within a herd over long distances.

DINOSAUR EGGS AND BABIES

Like birds and reptiles today, dinosaurs laid eggs to give birth to their babies. These eggs came in all shapes and sizes, from tiny eggs to giant ones the size of footballs.

Some were hard shelled like bird eggs, while others were soft shelled like those of turtles. Certain dinosaur eggs were even leathery like those of modern-day reptiles.

By uncovering clues about dinosaur reproduction and parenting, researchers can learn more about the lives of dinosaurs. Scientists also study dinosaur eggs and babies to find evolutionary clues that create a link between today and millions of years ago.

Nesting Sites

Dinosaur eggs have been found in a variety of places, from sandy beaches to rocky cliffs. Some dinosaurs built nests out of twigs, leaves, and mud to protect their eggs, while others simply laid their eggs in shallow depressions in the ground.

Fossilized nests with eggs arranged in a circular pattern suggest that some dinosaurs such as *Maiasaura* may have sat on their eggs like chickens.

Hatching and Birth

Baby dinosaurs hatched from their eggs by breaking through the shell using a sharp egg tooth. This tiny tooth, which is still present in some modern birds, helped baby dinosaurs crack open their eggs and emerge into the world.

Growing Up

Life wasn't easy for baby dinosaurs. Once they hatched, they had to fend for themselves and grow up quickly to avoid becoming lunch for hungry predators. Some baby dinosaurs, like *Triceratops* and *Tyrannosaurus*, were born ready to walk and even run shortly after hatching. Others, like *Hadrosaurus*, were more vulnerable and had to be cared for by their parents.

Examples of Baby Dinosaurs

Here's a closer look at some examples of baby dinosaurs based on scientific evidence and artistic interpretations.

Protoceratops: These small, horned dinosaurs may have laid their eggs in underground nests to protect them from predators. Fossilized nests containing eggs and hatchlings have been found, suggesting that *Protoceratops* parents may have guarded their nests and cared for their babies.

Brachiosaurus: Some of the largest dinosaurs such as *Brachiosaurus* laid eggs that were relatively small compared to their adult size. Babies had to grow quickly to keep up with their towering parents.

Maiasaura: Known as the "good mother lizard," *Maiasaura* likely cared for its eggs and young in nesting colonies. Fossilized nests containing eggs and young babies suggest that *Maiasaura* parents may have nurtured and protected their offspring after hatching.

DINOSAUR EXTINCTION THEORIES

Dinosaurs ruled the Earth for millions of years before suddenly disappearing. Scientists have been trying to solve this ancient puzzle for decades, and while we may not have all the answers, there are some intriguing theories to explore. We've briefly talked about some of these theories already, but let's take a little deeper look at each one.

Asteroid Impact

One of the most famous theories about dinosaur extinction suggests that a massive asteroid crashed into Earth around 66 million years ago. The impact caused wildfires, tsunamis, and a massive dust cloud that blocked out the sun.

Without sunlight, plants died, and there wasn't enough food for the dinosaurs to survive.

The asteroid impact theory offers a clear and dramatic explanation for the demise of the dinosaurs. It's supported by the presence of large craters such as the Chicxulub crater in Mexico.

Volcanic Activity

Another theory is that volcanic eruptions played a significant role in the extinction of the dinosaurs. Around the same time as the possible asteroid impact, massive volcanic eruptions occurred in an area known as the Deccan Traps in present-day India. These eruptions released vast amounts of lava, gasses, and ash into the atmosphere, leading to climate changes and environmental disruptions. The combination of volcanic activity and the asteroid impact may have been too much for dinosaurs to survive.

Climate Change

Dinosaurs lived in a world with changing climates. Some scientists think that long-term climate changes, such as cooling temperatures or changing sea levels, may have contributed to dinosaur extinction. These changes could have affected habitats and food sources, making it difficult for dinosaurs to survive and reproduce.

Disease

Dinosaurs could have gotten sick from viruses, bacteria, or parasites. If too many dinosaurs lived close together, or if their environment drastically changed, it might have made them more likely to catch diseases. New germs could also have come from migrating dinosaurs and made the local ones even sicker. When dinosaurs got sick, it could have made it hard for them to have babies and stay healthy, which might have led to some dinosaurs disappearing forever.

How Long Did It Take for Dinosaurs to Go Extinct?

The event that caused dinosaur extinction happened relatively quickly. Scientists estimate that it took anywhere from a few thousand to a few million years for dinosaurs to go extinct after the catastrophic event. While some dinosaur species disappeared immediately, others may have lingered for a while before finally disappearing.

The extinction of dinosaurs is one of the greatest mysteries of our planet's history. While we may never know for sure what caused their demise, scientists continue to study fossils, rocks, and other clues to piece together the events that led to the end of the dinosaur age.

CHAPTER SIX: INTERACTIVE ACTIVITIES

"DID YOU KNOW" FACTS

Dinosaurs continue to captivate our imaginations and probably will for future generations to come. Now that you are familiar with the basics about these prehistoric giants, let's explore some fascinating facts that often escape common knowledge. You can share these facts with your friends or turn them into a pop quiz.

Dino Symphony

Did you know that some scientists believe certain dinosaurs like *Parasaurolophus* could produce low-frequency sounds like those made by modern-day elephants? These resonating calls could have been used to communicate over long distances.

Night-Vision Predators

Did you know that many dinosaurs were nocturnal? Contrary to popular belief, not all dinosaurs were active during the day. Some, like the fearsome *Velociraptor*, may have been nocturnal hunters. Their large eyes and keen senses could have given them the advantage of hunting under the cover of darkness.

Dino Dentistry

Did you know that dinosaurs had a variety of dental quirks? For instance, the *Diplodocus* had pencil-like teeth, while the *Hadrosaurus* had rows of teeth perfect for grinding tough vegetation. Some dinosaurs even shed and replaced their teeth throughout their lives, much like modern sharks.

Some dinosaurs may even have practiced dental hygiene. Fossilized tooth remains of certain species show patterns consistent with picking at their teeth using their claws. This suggests that they might have been cleaning their teeth.

Dino DJs

Did you know that the hollow crests on the heads of dinosaurs like *Corythosaurus* and

Lambeosaurus weren't just for show? Scientists speculate that these cavities could have been used as resonating chambers, allowing them to produce unique calls or even communicate through music-like sounds.

Dino Preening

Did you know that dinosaurs with feathers likely engaged in preening behaviors to keep their feathers clean and in good condition? This grooming ritual may have been crucial for staying warm and displaying vibrant colors while searching for a mate.

Dino Artists

Did you know that the rocks with preserved dinosaur footprints occasionally contain scratch marks that look like they were purposefully left behind? These markings could have been a way of communicating or marking territory.

DINOSAUR KITS AND GAMES

While visiting museums and reading books are excellent ways to learn more about dinosaurs, you can create your own dinosaur adventures right at home. These activities can give you insight into the ancient world. You may even be able to convince your friends or family members to join in.

Excavation Kits and Models

If you enjoy hands-on activities, Discovery #Mindblown fossil kits let you excavate, assemble, and display models of different dinosaurs. The pieces of the model are hidden inside a block of clay. You'll use the included tools to uncover each piece and figure out how they fit together. One kit contains *Tyrannosaurus*, while the other features *Velociraptor*.

Card Games

Card games such as *Guess in 10: Deadly Dinosaurs* allow you to test your knowledge while having fun with loved ones. One player draws a card but doesn't show it to anyone else. Another player can ask up to 10 questions about the dinosaur on the card before making a guess. The deck includes familiar dinosaurs such as *Archaeopteryx* as well as others you might not recognize.

Dino Arts and Crafts

Get creative with dino-themed arts and crafts projects that will bring the world of dinosaurs to life. Design your own dinosaur masks, make dinosaur footprints, or create colorful dinosaur paintings and sculptures.

Dino Movie Night

Host a back-yard movie night featuring your favorite dinosaur movies and documentaries. Set up a projector or a laptop outdoors, hang up some blankets or set up lawn chairs, and get cozy under the stars. From classic films like *Jurassic Park* to educational documentaries, there's no shortage of dino-themed entertainment to enjoy with friends and family.

CHAPTER SEVEN: REAL-LIFE DINOSAUR ENCOUNTERS

DINOSAUR MUSEUMS AND EXHIBITS

It's fun to read about dinosaurs in books or see movies about these incredible animals. However, it's even better to get up close and personal with them at museums and exhibits around the world. There are dozens of places you can go to learn more about dinosaurs, but here are a few of the best.

American Museum of Natural History: New York City, NY

Located in the heart of New York City, the American Museum of Natural History features an incredible collection of dinosaur fossils. From the towering *Tyrannosaurus rex* to gentle giants like the *Apatosaurus*, you'll marvel at the size and diversity of these ancient creatures. Don't miss the famous T. rex fossil named Sue and the interactive exhibits that let you experience life as a paleontologist.

Royal Tyrrell Museum: Alberta, Canada

Nestled in the rugged badlands of Alberta, Canada, the Royal Tyrrell Museum is a paradise for dinosaur enthusiasts. Guests can explore lifelike exhibits with prehistoric landscapes and see some of the most well-preserved dinosaur specimens ever discovered. You might even get the chance to participate in a real fossil dig.

Natural History Museum: London, UK

In the heart of London, the Natural History Museum is home to one of the world's most famous dinosaur exhibits. Meet the iconic *Diplodocus* skeleton that's affectionately known as "Dippy," and learn about the incredible creatures that once ruled the Earth. With hands-on activities and multimedia displays, this museum offers a truly immersive experience for young paleontologists.

Fernbank Museum of Natural History: Atlanta, GA

Step into a prehistoric world at the Fernbank Museum of Natural History in Atlanta, Georgia. Marvel at the massive *Argentinosaurus* skeleton, explore ancient habitats, and even dig for fossils in the outdoor quarry. With interactive displays and educational programs, this museum is perfect for budding dinosaur enthusiasts.

National Dinosaur Museum: Canberra, Australia

Venture down under to the National Dinosaur Museum in Canberra, Australia, where you'll encounter a diverse selection of dinosaur fossils and exhibits. From the fearsome *Allosaurus* to the peaceful *Stegosaurus*, you'll discover the wonders of prehistoric life in the southern hemisphere. Be sure to check out the museum's collection of dinosaur eggs and learn how scientists study these ancient artifacts.

Zigong Dinosaur Museum: Sichuan, China

Travel to the heart of Sichuan Province in China to visit the Zigong Dinosaur Museum, one of the largest dinosaur museums in the world. Featuring stunning displays of dinosaur skeletons, fossils, and animatronic models, this museum offers a thrilling glimpse into China's rich paleontological history. Explore the museum's outdoor park to see life-sized replicas of various dinosaur species in their natural habitats.

Bernardino Rivadavia Natural Science Museum: Buenos Aires, Argentina

Embark on a dinosaur adventure in the heart of Buenos Aires at the Bernardino Rivadavia Natural Science Museum. Discover the incredible diversity of South American dinosaurs, including the massive *Giganotosaurus* and the unique herbivore *Amargasaurus*. With interactive exhibits and educational programs, this museum is a must-visit for young explorers fascinated by dinosaurs.

DINOSAUR FOSSIL PARKS

One of the best ways to experience dinosaurs is by visiting dinosaur fossil parks around the world. These parks are treasure troves filled with clues about the prehistoric past. Museums are wonderful, but if you want a hands-on experience, try one of these parks instead.

Dinosaur National Monument: Utah and Colorado

Our first stop is in the United States, where you can visit Dinosaur National Monument. Located in Utah and Colorado, this park is famous for its incredible dinosaur fossils. Imagine hiking through rugged canyons and stumbling upon the bones of long-extinct dinosaurs. You might even get to see a real dinosaur quarry where scientists carefully unearth fossils.

Dinosaur Provincial Park: Alberta, Canada

This UNESCO World Heritage Site is like stepping into a prehistoric paradise. Explore vast badlands and picturesque landscapes while keeping an eye out for fossils hidden amongst the rocks.

Joggins Fossil Cliffs: Nova Scotia, Canada

This coastal gem is famous for its well-preserved fossils dating back over 300 million years. Walk along the beach and uncover fossils of ancient plants, insects, and even early reptiles. It's like taking a journey through the evolution of life on Earth.

Dinosaur Valley State Park: Glen Rose, TX

Here, you'll be able to see real dinosaur footprints preserved in ancient riverbeds. Walk in the footsteps of dinosaurs as you explore the park's scenic trails and rugged terrain. There are five separate track sites within the park.

Ischigualasto Provincial Park: San Juan Province, Argentina

Also known as the Valley of the Moon, this UNESCO World Heritage Site features remarkable dinosaur fossils and landscapes. You can explore the park's unique rock formations while immersing yourself in the ssbeautiful scenery. Ischigualasto Provincial Park promises an unforgettable journey into the distant past.

DINOSAUR TRACKSITES AROUND THE WORLD

Not only did dinosaurs leave behind fossils, but they also left footprints preserved in rocks. These are known as *tracksites*. There are dinosaur tracksuits on every continent in the world — even Antarctica!

United States

In the vast expanses of the United States, iconic dinosaur tracksites like Dinosaur Valley State Park in Texas and Dinosaur Ridge in Colorado showcase footprints left by formidable dinosaurs such as *Tyrannosaurus* and *Brachiosaurus*. Walking in their footsteps, you can almost feel the ground tremble beneath you, transporting you back to an era when these magnificent creatures roamed the land.

Australia

Journey to the Dinosaur Stampede National Monument where preserved footprints depict a scene of dinosaur chaos. Hundreds of small footprints were caused by a stampede, offering a unique glimpse into the behaviors of these ancient beasts.

South America

Bolivia's Cal Orck'o tracksite is a marvel of paleontological discovery. With its vast expanse of over 5,000 dinosaur footprints, it holds the Guinness World Record for the largest collection. These footprints, etched into the landscape millions of years ago, offer a snapshot of ancient life on Earth. The site provides scientists with invaluable data on how dinosaurs behaved, moved, and interacted.

Visitors to Cal Orck'o can witness these incredible tracks up close, marveling at the scale and diversity of life that once flourished in this region. It's a journey through time that ignites curiosity and wonder in all who visit.

Europe

Journey to the Isle of Skye in Scotland where ancient mudflats preserve the footprints of dinosaurs. These tracks offer a window into the lives of prehistoric creatures that once roamed this area.

Africa

Tanzania's Tendaguru Beds are famous for their spectacular dinosaur fossils and tracks. As you wander among these ancient footprints, you'll be transported to a time when dinosaurs roamed the African savannahs.

Asia

In China, the Zhucheng Dinosaur Museum stands as a testament to the region's rich paleontological history. Inside, visitors are treated to an awe-inspiring collection of dinosaur tracks that have been preserved and displayed for all to see. These tracks offer valuable insights into the behavior and movements of various dinosaur species that once roamed the land.

Antarctica

In the icy expanse of Antarctica lies a surprising treasure: a tracksite revealing the presence of dinosaurs from over 200 million years ago! Near the Beardmore Glacier, fossilized footprints of sauropods and theropods offer insights into their ancient lives. Despite the harsh conditions, scientists brave the cold to unravel Antarctica's prehistoric mysteries, studying how dinosaurs adapted to polar environments.

CHAPTER EIGHT: DINO CONSERVATION

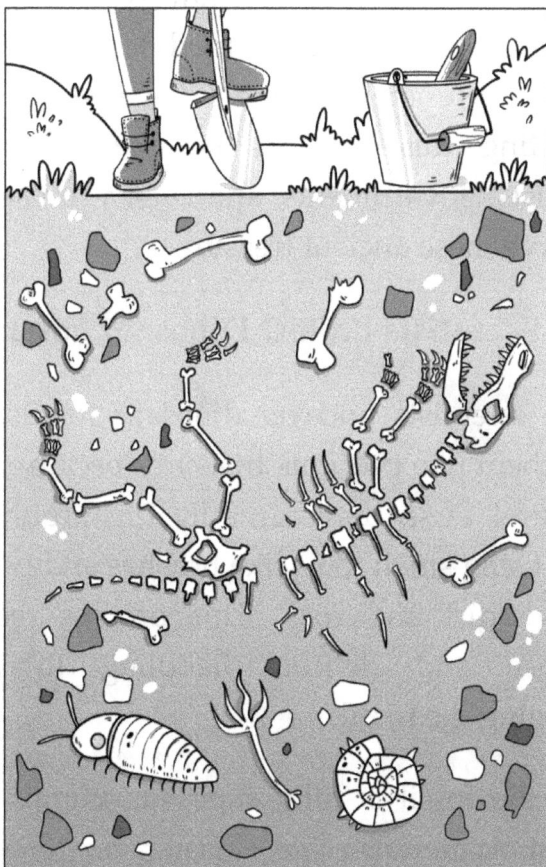

PROTECTING DINOSAUR FOSSILS

Protecting dinosaur fossils is just as important as finding them. However, how do we protect something that's been buried for millions of years? Fortunately, experts have spent years perfecting this process. Let's look at how scientists, governments, and even you can help preserve these ancient treasures.

How Scientists Protect Dinosaur Fossils

When scientists uncover dinosaur fossils, they treat them like precious treasures because, well, they are! First, they carefully dig around the fossil using small tools like brushes and picks to remove dirt and rock. This process requires patience and delicate handling to avoid damaging the fossil.

Once the fossil is fully exposed, scientists use special materials like plaster to create a protective jacket around it. This jacket keeps the fossil safe during transport to a museum or research facility. At the museum, experts clean the fossil further and study it to learn more about the ancient creatures that once roamed the Earth.

Scientists also document everything they find near the fossil, like plants and rocks, to understand the environment the dinosaur lived in. This helps them piece together the puzzle of what life was like millions of years ago.

How Governments Protect Dinosaur Fossils

Governments play a crucial role in safeguarding dinosaur fossils. Many countries have laws in place to protect these valuable pieces of history. For example, it's illegal to remove fossils from certain areas without permission. Governments also establish national parks and reserves where fossils are protected, allowing scientists to study them while keeping them safe.

Some governments even have special teams like park rangers who patrol fossil sites to ensure they remain undisturbed. These efforts help preserve fossils for future generations to enjoy and learn from. For example, in the United States, the Bureau of Land Management oversees the protection of dinosaur fossils in places like Dinosaur National Monument in Utah and Colorado.

Rangers patrol these areas to prevent criminal activity and illegal fossil removal. Additionally, strict regulations are in place to ensure that fossils remain undisturbed, allowing scientists to study them while preserving them for future generations.

How Kids Can Help Protect Dinosaur Fossils

You might be wondering, "How can I help protect dinosaur fossils as a kid?" Well, there are plenty of ways you can contribute to preserving these ancient treasures.

First and foremost, always remember to respect fossil sites. If you ever come across a dinosaur bone or footprint while exploring, resist the urge to touch or remove it. Instead, take a picture and let an adult know so that they can report the discovery to authorities.

You can also spread awareness about the importance of protecting fossils among your friends and family. If more people understand why fossils should be left undisturbed, we'll have a better chance of preserving them for future generations.

Lastly, you can support organizations and museums that work to protect and study dinosaur fossils. Whether it's visiting a museum exhibit or donating to a conservation program, every little bit helps.

By working together to protect these amazing fossils, we can ensure that their stories continue to inspire and educate people of all ages for years to come. So let's join forces and become guardians of the past, preserving these ancient treasures for future generations.

PRESERVING DINOSAUR HABITATS

For many years, humans didn't really pay much attention to dinosaur habitats. However, in recent decades, society has recognized the importance of maintaining biodiversity, protecting ecosystems, and learning from Earth's ancient history. By safeguarding these habitats, we can ensure the survival of diverse species and gain valuable insights into the past.

How Scientists Preserve Dinosaur Habitats

Scientists study fossils, rocks, and plants to piece together the ecosystems where dinosaurs roamed. However, preserving dinosaur habitats involves more than just studying them. Preservation is also about ensuring these environments remain intact for future generations to explore.

One way that scientists preserve dinosaur habitats is by conducting research to understand the needs of modern-day animals living in similar environments. By studying the behavior and habitats of birds and reptiles, scientists can make informed decisions about how to protect dinosaur habitats and the creatures that now call them home.

Another important aspect of habitat preservation is conservation efforts. Scientists work with governments and conservation organizations to establish protected areas such as national parks and reserves where dinosaur habitats are shielded from harm. These protected areas provide safe havens for plants and animals, ensuring that the ecosystems remain healthy and thriving.

How Governments Preserve Dinosaur Habitats

Governments play a vital role in safeguarding dinosaur habitats through legislation and conservation initiatives. Many countries have laws in place to protect natural areas and prevent habitat destruction. For example, it's illegal to build on or disturb certain habitats where dinosaur fossils are found without proper authorization.

Governments also establish national parks and wildlife reserves to preserve habitats and other creatures. These protected areas offer safe havens where plants and animals can thrive without the threat of human interference. Park rangers and conservation officers patrol these areas to ensure that habitats remain undisturbed and wildlife can flourish.

How Kids Can Help Preserve Dinosaur Habitats

There are plenty of ways you can contribute to protecting ancient dinosaur habitats. One simple way to help is by learning about the plants and animals that live in dinosaur habitats and spreading awareness about their importance.

You can also get involved in local conservation efforts by volunteering with environmental organizations or participating in community clean-up events. By taking action to protect the environment in your own back yard, you're helping to preserve habitats for countless other species.

Lastly, you can make environmentally friendly choices in your daily life, such as reducing waste, conserving water, and recycling. Every small step you take to live more sustainably makes a difference in preserving habitats for all living things on Earth.

CONCLUSION

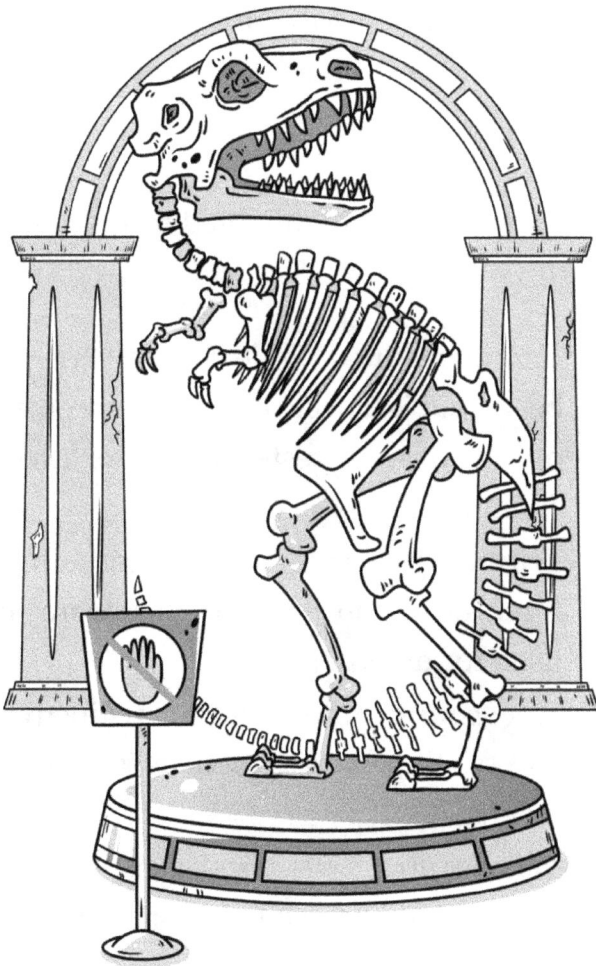

RECAP OF AWESOME DINOSAUR FACTS

In this book, we've covered a wide range of fascinating facts about various dinosaurs, providing insight into their unique characteristics, behaviors, and adaptations. Let's recap the key points highlighted throughout our journey.

- *Tyrannosaurus*: While its small arms have been a subject of curiosity and speculation, one theory suggests they were used for swimming.

- *Triceratops*: This iconic dinosaur had three horns on its head that were likely used for defense.

- *Stegosaurus*: Known for its distinctive tail spikes, *Stegosaurus* used its tail for self-defense.

- *Velociraptor*: Its sickle claw was a formidable weapon used for slashing prey and hunting.

- *Brachiosaurus*: This sauropod dinosaur had extra chambers in its heart, a unique adaptation that likely helped it maintain healthy blood pressure and circulation.

- *Ankylosaurus*: Its tail club was a key defensive feature, providing protection against predators through powerful strikes.

- *Parasaurolophus*: The hollow bony crest on its head served as a resonating chamber, possibly used for communication.

- *Spinosaurus*: Distinguished by the sail-like structure on its back, *Spinosaurus* was a unique predator adapted to both land and water environments.

- *Allosaurus*: Known as a fierce hunter, *Allosaurus* was a formidable predator in the Late Jurassic period.

- *Diplodocus*: Its incredibly long neck allowed *Diplodocus* to reach high into the treetops to feed on vegetation.

- ***Archaeopteryx***: An early bird-like dinosaur, *Archaeopteryx* had both feathers and wings.

- ***Iguanodon***: This herbivorous dinosaur was originally thought to be an ancient type of iguana.

- ***Carnotaurus***: Large, forward-facing eyes likely gave *Carnotaurus* enhanced vision and depth perception to help it hunt.

- ***Microraptor***: Feathers on all four limbs made it easier for *Microraptor* to easily glide through the air.

- ***Apatosaurus***: Initially mistaken for a different dinosaur, the name *Apatosaurus* means "deceptive lizard" because scientists were often confused about its fossils.

- ***Utahraptor***: Reaching up to 1,000 pounds, *Utahraptor* is the largest known raptor.

- ***Edmontosaurus***: Its duck-like bill helped this herbivore grind up tough vegetation.

- ***Brontosaurus***: *Brontosaurus* might have used its strong neck to fight and defend itself.

- *Gallimimus*: *Gallimimus* could run as fast as 34 miles per hour.

- *Stegoceras*: Scientists aren't sure whether the bony dome on its skull was used for fighting or defense.

- *Oviraptor*: This omnivorous raptor probably didn't eat eggs at all even though its name means "egg thief."

- *Deinonychus*: With a longer tail than *Velociraptor*, *Deinonychus* had unique physical features and hunting behaviors.

- *Compsognathus*: Only two sets of *Comsognathus* fossils have ever been found.

- *Protoceratops*: *Protoceratops* had a horned beak instead of horns above its eyes or on its nose like *Triceratops*.

- *Maiasaura*: Known as the "good mother lizard," *Maiasaura* exhibited remarkable nesting and parenting behaviors, caring for its young in a manner like modern birds.

EMBRACING THE WONDER OF DINOSAURS

As we come to the end of this book, we're sure you're even more curious about dinosaurs than ever before! Fortunately, we want you to embrace the wonder of dinosaurs and continue learning as much about them as you can. We've put together some ideas for how you can keep your curiosity alive.

First, keep in mind that there are hundreds upon hundreds of books about dinosaurs. Some have real pictures of dinosaur fossils, while others include illustrations of what dinosaurs might have looked like when they were alive. Grab another book about dinosaurs from the library or bookstore and let your imagination run wild.

Next, consider a trip to one of the museums and exhibits mentioned earlier. Picture yourself standing face to face with real dinosaur fossils, feeling the thrill of discovering these ancient creatures up close. Many museums have awesome exhibits designed just for kids, with

122

interactive displays and hands-on activities that make learning about dinosaurs even more fun.

How about getting your hands dirty with some dinosaur-themed activities? You can create your own fossils, dig for dinosaur bones in an activity kit, or even make your own dinosaur habitat. These DIY projects are both educational and fun.

Let's not forget about technology. With documentaries, videos, apps, and websites, the world of dinosaurs is right at your fingertips. You can watch *Tyrannosaurus* come to life on the screen, play dino-themed games, and even take virtual tours of dinosaur digs.

So why stop there? Learning about dinosaurs isn't just about memorizing facts — it's also about sparking your imagination and curiosity about the world around you. Why not turn your passion for dinosaurs into a career? Studying dinosaurs can teach us important lessons about the environment and conservation. Of course, paleontologists have the most well-known dinosaur-related profession, but here are some others:

- **Paleoartist:** Paleoartists use their artistic skills to create reconstructions of dinosaurs and other prehistoric creatures based on fossil evidence. They work closely with paleontologists to accurately depict what dinosaurs may have looked like, bringing these ancient animals to life through illustrations, sculptures, and animations.

- **Museum curator:** Museum curators are responsible for managing and preserving collections of dinosaur fossils and other artifacts. They oversee the acquisition and display of specimens, ensuring that they are properly cared for and presented to the public.

- **Educator:** Educators who specialize in paleontology teach students of all ages about dinosaurs and the science of paleontology. They may work in schools, museums, or other educational institutions to inspire the next generation of dinosaur enthusiasts.

- **Paleoecologist:** Paleoecologists study the ancient environments in which dinosaurs lived, including the plants, animals, and ecosystems of the past. By analyzing fossilized plants, pollen, and other evidence, they can reconstruct the habitats and ecosystems of the Mesozoic Era. This allows paleoecologists to understand the roles of dinosaurs and how they interacted with their environment.

- **Geologist:** Geologists often work alongside paleontologists to understand the geological processes that preserved dinosaur fossils. They study the rock layers where fossils are found, using techniques like stratigraphy and radiometric dating to determine the age of natural features.

- **Conservator:** Conservators specialize in the preservation and restoration of artifacts, including dinosaur fossils. They use specialized techniques and materials to stabilize, repair, and protect fossils, ensuring that they remain intact and accessible for future generations to study and enjoy.

Whether you're flipping through books, exploring museums, getting creative with hands-on activities, diving into the digital world, or studying for a career, there's so much more to discover and learn about dinosaurs.